Welwitcha Sory dos Santos Lima

Simulation of the process of obtaining Maltina

Welwitcha Sory dos Santos Lima

Simulation of the process of obtaining Maltina

from red sorghum CIAP R--132 a132 at pilot plant scale

ScienciaScripts

Cover image: www.ingimage.com

This book is a translation from the original published under ISBN 978-620-2-14069-0.

Publisher:
Sciencia Scripts
is a trademark of
Dodo Books Indian Ocean Ltd. and OmniScriptum S.R.L publishing group

120 High Road, East Finchley, London, N2 9ED, United Kingdom
Str. Armeneasca 28/1, office 1, Chisinau MD-2012, Republic of Moldova, Europe
Printed at: see last page
ISBN: 978-620-7-70559-7

Contents

Thinking

"Education is the most powerful weapon you can use to change the world" Nelson Mandela.

Dedication

To God for having guided and protected me here. To my parents for supporting me to make this dream possible. To my sisters, that one day they will give me the satisfaction of seeing them defend this job.

Acknowledgement

Firstly, to thank God for the gift of life. To my parents for their immense love and dedication, especially to my mother for being a tireless woman.

To my tutors Isnel Benitez and Amaury Sanchez for all their help, dedication and patience. To my friends and classmates, especially my baby Galcione Felicidade Chitali who in good times and bad have been by my side. To all the teachers who taught me during these beautiful five years.

Summary

The simulation of a pilot scale maltine production plant (150 L/batch) was carried out, using red sorghum CIAP R-132 COTO as the main raw material. A sensitivity study consisting of 11 experimental runs was carried out to evaluate the influence of three input variables (maltine production capacity per batch; cost of purchasing red sorghum; and selling price of the maltine bottle) on three important economic indicators: Net Present Value (NPV), Internal Rate of Return (IRR) and Investment Recovery Period (IRP). The unit cost of production reached a value of $ 3.73/bottle, while the NPV, IRR and IRR of $ 3,661,000, 63.20% and 1.02 years, respectively, were obtained, which qualifies the process as economically profitable and feasible from the investment point of view. The sensitivity study allowed to obtain equations that establish the statistical correlation between the 3 input variables and the 3 output variables. The plant should have a minimum production capacity of 125 L/batch, a sorghum purchase cost of $ 16.39/kg and a maltine bottle selling price of $ 5 to obtain a positive NPV value ($ 424,356). Finally, the quantification of the emissions of solid, liquid and gaseous residues obtained in the production process was carried out. The SuperPro Designer® v.8.5 simulator was used for the simulation and the Statgraphics Centurion XVI® software was used for the statistical processing of the data. Keywords: Maltine; sorghum; simulation; sensitivity analysis; profitability.

Keywords: Maltina; sorghum; simulation; sensitivity analysis, profitability.

Introduction

At present, the chemical industry сото mission is to base its development on research into products that can be attractive from the point of view of their use, quality, market, which would also lead to their technical feasibility, economic and environmental sustainability.

It is for this reason that the development of natural products in the pharmaceutical, food, cosmetics, dyes and biotechnological industries is becoming more and more important.

The use of cereals for animal feed has been a dynamic element in global sorghum consumption. Demand for sorghum has been the main driving force in increasing world production and international trade since the 1960s. Sorghum is the fifth most important cereal in the world, after wheat, rice, maize and oats (Pacheco, 1998). The main sorghum production sites are in the arid and semi-arid regions of the tropics and subtropics (Doggett, 1998). In Africa, a significant part is destined for human consumption, while in America and Oceania most of the sorghum produced is used for animal consumption, for example, in cattle feed (Ostrowski, 1998) (Salermo, 1998), in poultry, as well as being widely used in other countries as a raw material in the starch and alcohol and brewing industry (Lyumugabe et al., 2012) (Ramatoulaye et al., 2016).

Sorghum demand is heavily concentrated in countries such as the United States of America, with a production of 11.9 million tonnes (Mt) of grain, India (9.5 Mt), Nigeria (7.5 Mt) and Mexico (6.4 Mt), which are considered leading producers (Perez et al., 2010).

There is currently great interest in the use of sorghum as a source of energy in the production of food for human consumption, replacing corn in the production of food concentrates (Valencia & Rooney, 2009). Moreover, its chemical composition is practically identical to that of corn, so it can be used instead to produce flour, tortillas, starch, syrups and alcoholic beverages. Without

However, it has the disadvantages of having a peripheral endosperm that acts as a barrier against the penetration of the soaking solution, a harder, cross-linked protein matrix surrounding the starch granules, lowering starch yields and quality (Rooney & Serna, 2000), as well as starch recovery during wet milling (Serna, 1998).

Barley malt has traditionally been used in the production of beverages, сото beer y maltinas y in the direct production of ethanol, for its diastatic o amylolytic activity, compared with that of other cereals, but for many countries сото Cuba is a raw material of import, despite this there are people who can not eat the food produced from it by the type of protein that presents this cereal (Rodriguez et al., 2015). However, sorghum is a cereal that contains many beneficial properties and it has been shown that its cultivation is economically profitable, based on its low cost of production, given its characteristics of hardiness, resistance to drought, the realization of several harvests or cuts, etc., and does not contain the proteins that affect celiac patients (Gallardo et al., 2013). The development of products from sorghum, including maltine, has aroused great interest in the region (Diaz, 2014). Process simulation for the evaluation of alternatives and loss mitigation in a given process will play a very important role for the final optimisation model, as it can reveal on a large scale whether a future process can give successful results or not be highly feasible. The simulation of a chemical plant consists of the creation of a process model, where a model is understood as a description of the behaviour of a real process, capable of predicting the output (responses) as a function of the inputs.

Celiac disease, or gluten-sensitive enteropathy, is an intestinal disorder with a multifactorial aetiology in which a malabsorption syndrome develops, among other complications, produced by damage to the intestinal villi of the small intestine when gluten is ingested (Carvajal, 2014). This process occurs in genetically predisposed individuals after ingestion of some of the proteins present in cereals (gluten), especially *prolamins*, which are named differently depending on the cereal: *gliadins* in wheat; *hordeins* in barley; *secalins* in rye y although not so clearly demonstrated the *avenin* in oats (Pino, 2017).

The *SuperPro Designer^* software (Intelligen, 2014) allows the simulation of various processes and operations on an industrial scale using a computer. In many fields, such as chemical, biotechnological and petroleum, specific simulators have been developed, including Aspen Plus®, ChemCAD® and Hysys®. Among the many existing simulators, *SuperPro Designed* is one of the most relevant for the chemical process industry, which has been used mainly for mass balances, feasibility studies, economic analysis of variants, as well as design and conceptual evaluation of industrial processes and systems. Currently, 18 of the top 20 biopharmaceutical companies worldwide use this software for their daily operations (Intelligen, 2014).

Sensitivity analysis is a tool used to estimate the contribution of *input* factors to *output* variables. In this case, the aim is to determine the possible variation of the initial outlay, the cash flows and the discount rate in order to make the investment worthwhile. This analysis is carried out according to the NPV y according to the IRR. The sensitivity analysis to determine the profitability of an investment, it can be said that it is a financial analysis that is essential to measure the performance of the business and to identify the problems that may arise.

Red sorghum variety CIAP R-132 has been used to evaluate the wet milling process for the extraction of starch contained in the grains (Rodriguez et al., 2015); the production of dextrinised syrups by enzymatic hydrolysis of starch applying the enzyme a-amylase

(Rodriguez et al., 2015); сото adjunct component blended with barley for obtaining beer y malt at pilot scale (50 L) (Nieblas et al., 2016) y сото sole raw material for obtaining beer at laboratory scale (Alfonso, 2018). However, to date, it has not been reported to be used in a comprehensive way for maltine production at pilot plant scale, so it is not known what values will have the main technical-economic indicators of such a plant.

The elements described so far show the need for a study to analyse the economic feasibility of a maltine production plant on a pilot scale (150 L/batch) using red sorghum CIAP R-132 сото integral raw material, therefore, the proposed **research problem** is сото:

Is the proposal of a technological plant for the production of maltine from red sorghum CIAP R-132 on a pilot scale feasible in the current economic context of Cuba?

Hypothesis:

By simulating the proposed technology for maltine production from red sorghum CIAP R-132, the values of its main technical and economic indicators can be determined in order to evaluate its profitability and yield.

General objective:

Simulate the maltine production plant from red sorghum using SuperPro Designer to determine the technical-economic indicators that will allow the feasibility of the process to be assessed on a pilot scale.

Once the general objective of the work has been set out, the following **specific objectives are** presented:

1. To carry out a bibliographic study on the production processes of maltine from sorghum and its most important characteristics.
2. Simulate the proposed production technology using the *SuperPro Designed* simulator.
3. Analyse the values of the main technical-economic indicators obtained by the simulator for the proposed production technology.
4. Conduct a sensitivity study containing 3 input and 3 output variables, and evaluate the statistical correlation between them.

After setting out the specific objectives, the **tasks** to be carried out are shown:

1. To carry out a bibliographic study on the subject studied (production of maltine from sorghum).
2. To simulate the maltine production process from red sorghum CIAP R-132.
3. Analyse the technical and economic results obtained and compare them with the existing literature on the subject.
4. Conduct a sensitivity study containing 3 input variables and 3 output parameters.
5. Determine the statistical correlation between the 3 initial variables considered in the sensitivity study, on 3 important economic parameters of the process (NPV, IRR y PRI).

CHAPTER 1

State of the Art

1.1. What is Maltina?

Malt is a type of soft drink. It is a carbonated malt beverage, i.e. it is prepared from barley, hops and water, and water, and may also contain corn and caramel-coloured beer. However, malt does not contain alcohol and is consumed in the same way as soda or cola in its original carbonated form and, to some extent, сото ice tea in non-carbonated form.

In other words, malt is a beer that has not been fermented. It is similar in colour to stout (dark brown) but is very sweet, usually said to taste like molasses. Unlike beer, ice is often added to malt when it is consumed. Malt originated in Germany сото Malzbier ("malt beer"), a dark malty beverage, whose fermentation stops at about 2% ABV, leaving a large amount of residual sugars in the finished beer. In the 1950s, Malzbier was considered a fortifying food for nursing mothers, recovering patients, the elderly, etc. (Leyanis Rodriguez, Gallardo, Nieblas, & Ortiz, 2015).

Malzbier in its original form was replaced in the 1960s by its modern form, formulated with water, glucose syrup, malt extract and molasses extract, which had been on the market since the second half of the 19th century, especially in Denmark. Such formulated beverages are called Vitamalz ('malt beer') under German law, since they are not fermented. However, in colloquial usage, Malzbier has remained alongside other nicknames сото Kinderbier.

Malt is sometimes referred to as "champagne cola" by some brands. However, there is another type of drink with this name, which has a taste and consistency more similar to cream soda. Despite this commercial designation, neither is champagne o cola. Because of its distinctive colour, malt is sometimes called сото "black beer" although in Chile malt is a fermented beer with alcohol, сото for example are the malts Morenita y del Sur. That is why in the latter country it is called Caribbean malt to avoid confusion.

Malt is rich in vitamin B. Some breweries, such as сото Albani Brewery in Denmark, fortify their malts with vitamin B complex. The Albani Brewery claims on its website that it was the first brewery to create non-alcoholic malt beverages in 1859. Malt is readily available in Latin America. It is more difficult to find in the United States, Mexico and Canada.

It should be noted that the malt, being made from barley, contains gluten, which makes it unsuitable for coeliacs o people with NCGS.

Today, most malt is brewed in the Caribbean and is available in areas with significant Caribbean populations. In addition to the Caribbean islands, malt is also popular in countries with Caribbean coastlines, such as Colombia, Panama and Venezuela, as well as in countries that share a Caribbean coastline. Malt is brewed all over the world, and is popular in many parts of Africa сото Nigeria, Chad, Ghana, Cameroon, and the Indian Ocean. It is also popular in some parts of Europe, especially Germany (Leyanis Rodriguez et al., 2015).

1.2. Benefitsofamaltine

Maltine has the following beneficial aspects (Ozuna, 2008) (Y. Diaz, 2014): J It is an energy drink, used by athletes because it provides quick energy and also provides proteins, vitamins, minerals and amino acids, which makes it the healthiest drink.

J *It* is suitable for children as it provides a number of essential nutrients for the development of bones and teeth.

J It acts as a hormone regulator in both men and women.

J The vitamins of the vitamin complex contained in them improve and revitalise the skin, hair and nails.

J Those suffering from anaemia may consume it as it contains iron and vitamin B.

J Contains a substance called Hamada inostol which regulates the body's fat metabolism by regulating cholesterol levels.

J *It* also acts on the capillaries, improving blood flow and preventing blockages.

J Drinking maltina is good for the heart.

J Due to its high amount of nutrients it is recommended to accelerate recovery after operations or illnesses.

J Contributes to bone remineralisation and is highly recommended to prevent osteoporosis.

1.3. Differences between sorghum malt and barley malt

If beverages are to be made from sorghum to fully replace barley, the contrasts between the malts of the two cereals must first be understood. In this regard, (Lyumugabe et al., 2012) summarises the most important physiological differences between them, which are shown below.

The first difference is the enzymes that degrade the endosperm. During germination, the hormone gibberellic acid, at low concentrations (0.1-0.2 ppm) induces the aleurone layer of barley to produce enzymes сото a-amylase, protease, pentosanases y endo-p-glucanase that degrade the endosperm, but this hormone does not play such a role in enzyme development in sorghum. For example, in sorghum, a-amylase and carboxypeptidases are produced by the scutellum, while endo-p-glucanase, limiting dextrinase and endo-protease grow in the starchy endosperm. In barley malting, on the other hand, a-amylase, endo-protease, dextrinase y endo-P-glucanase grow in the aleurone layer, while carboxypeptidases y p-amylase are found in the starchy endosperm.

The second contradiction is related to the characteristics of the endosperm. The endosperm of malted sorghum retains the compactness of starch y is not crumbled сото in barley malt grains (Palmer et al., 1989). The third contrast lies in the level of enzymes in the kernels. Sorghum malt kernels contain low levels of endo-p-glucanase y p-amylase (Aisien and Palmer, 1983). The fourth inconsistency lies in the losses during malting. Malting losses (respiration and root casting) of sorghum malt are around 20%, while malting losses of barley malt is 7% after 6 days of growth at 25°C and 25 °C (M. O. Diaz, 2016; Y. D. Rodriguez, 2014).

1.4. Componentsofamaltine

Maltine has the following components (Ozuna, 2008) (Carvajal, 2014) (Leyanis Rodriguez et al., 2015):

1.4.1. Green malt y caramel malt

Brewed under a special drying and roasting process. Malted grains develop the enzymes needed to convert the starch in the grain into sugar. Caramel malts are produced from malts of normal protein content. Steeping and germination are very intensive, resulting in high enzyme activity in green malt. During saccharification at 65 - 75 °C sugars are formed by the degradation of proteins into starch and amino acids. Both products react with each other in the roasting stage to form the characteristic colour and aroma.

These sugars are caramelised into longer chains which are not converted into simple sugars by enzymes during mashing. They act сото antioxidant by slowing down the oxidation process in the container, thus favouring the stability of the beverage and the denaturation of the proteins favouring the formation of foam in the maltine (Leyanis Rodriguez et al., 2015).

1.4.2. Lupulo

Hop is an alternative ingredient in the production of maltine and has no substitute. The hops are indispensable because of their pleasant bitter taste and their characteristic mild aroma, and they also contribute to a better keeping quality and to a longer lasting head. The hop is a wild climbing plant which, through careful cultivation over the centuries, has developed characteristics which give the product a special aroma and bitterness. Hop is cultivated in temperate and warm areas (Ratnavathi & Chavan, 2016). There are many different types of hops, due to the conditions of cultivation (soil and water), leading to differences in the quality of the products. The types of maltines and beers are usually related to certain types of hops. The addition of hops can be done in different ways, most often hop extract or ground hop grain is used.

1.4.3. Water

The characteristics of the brewing water have a great influence on the quality of the malt. Water is a basic element, it influences to a great extent the taste of the product, whether it is softer, stronger or sweeter, due to its different hardness and taste, and it contains minerals and salts in different proportions, depending on where it is obtained. Water can be obtained from the local network, from springs (both surface and underground), from lakes and rivers. The composition of water influences the production process. The purer the water is, the better the maltine taste will be controlled. If drinking water is used, its organoleptic characteristics must be completely normal (A. Perez et al., 2010).

1.5. Elsorgo

Sorghum (*Sorghum bicolor* L. Moench) is the fifth most widely consumed cereal in the world after wheat, rice, maize and oats. It is used in Africa for human food, while in America and Oceania it is mainly used for the production of flour and animal feed (Chaviano, 2005). It is a crop that has a wide geographical distribution due to its ecological plasticity, y its good agronomic behaviour gives it attributes that favour its consumption, as it is not a demanding crop to fertile soils, requires little cultivation and cleaning y it is one of the cultivated plants with the greatest resistance to drought (Leyanis Rodriguez et al., 2015).

The following table (Table 1) shows the composition of sorghum, and its comparison with different types of cereals most commonly used in the production of food and beverages (L. Rodriguez, 2010). As can be seen in this table, the composition of sorghum is very similar to that of other cereals, especially barley and wheat, which are in great demand in the modern food industry.

Table 1. Percentage composition of different cereals used in beverage production

Content	Matz	Rice	Sorghum	Wheat	Barley
Humidity	10,9	12,0	11,7	11,1	10,6
Starch	68,6	67,0	69,0	69,3	66,0
Protein	10,0	7,7	10,4	0,2	13,0
Grease	1,9	4,3	3,7	3,4	2,1
Fibre	3,4	2,3	1,9	2,2	5,6
Ashes	0,2	0,3	0,4	0,4	2,7
Extract	60,0	70,0	63,0	65,0	59,8
Nitrogenous materials	5,0	0,4	1,7	1,7	1,6
Cellulose	3,6	2,0	2,0	2,0	2,0

Sorghum grain contains mostly starch, a compound through which this cereal is valued for possible use in the beverage industry, as it is degraded to obtain sugars, which are then fermented to obtain alcoholic beverages. Approximately 80% of the cereal grain is composed mainly of carbohydrates, with starch being the largest proportion.

1.6. Sorghum varieties grown in Cuba

In Cuba, sorghum is widely accepted among small-scale cereal producers, due, among other economic reasons, to the introduction and generalisation of sorghum varieties that are better adapted to the different ecosystems where cereal production takes place, to the broad possibilities of including it in the annual rotation scheme, especially in conjunction with rice, and to the feasibility of using it in complementary production for both animal and human consumption (Diaz, 2014). In the country there are many varieties of sorghum, being the

UDG-110 the most cultivated due to its high use in the industry for the production of wholemeal bread, biscuits, gofio сото substitute for wheat flour, y in the manufacture of soft bread, biscuits, sweets y drinks partially replacing wheat flour y barley (Chaviano, 2005). Red sorghum variety CIAP R-132 is also cultivated, although to a lesser extent (Nieblas et al., 2016).

1.7. Sorghum-based beverages

1.7.1. Beer

It is known that beer has many variants with a wide range of shades due to the different brewing methods and ingredients used. It is generally amber in colour with shades ranging from golden yellow through reddish-brown to opaque black. It contains dissolved carbon dioxide (CO2) which manifests itself in the form of bubbles at ambient pressure, and is usually crowned with a more or less persistent foam, with a crystalline or cloudy appearance (Lyumugabe, Gros, Nzungize, Bajyana, & Thonart, 2012).

1.7.2. Ethanol

It can be obtained through two manufacturing processes: fermentation or decomposition of sugars contained in different fruits, and distillation, which consists of the purification of fermented beverages. In China, alcohol is produced from sorghum, where the alcoholic beverage industry is a major consumer of sorghum grain (L. Rodriguez, 2010).

1.7.3. Maltina

Name of a nutritious food drink obtained from a wort prepared from malted grains, previously subjected to a cooking process and flavoured with or without hop flowers. It differs from beer in that it is not subjected to fermentation, which is why it is considered a non-alcoholic beverage (Ozuna, 2008).

1.8. Characteristics and identity of maltine

The main characteristics that identify a malt are (Ozuna, 2008):

Colour: This is determined by the raw materials, especially the malt, which must be roasted, as the colour of the wort determines the colour of the product, together with the addition of caramel colour.

Foaming: must have a stable foam. The formation of the foam depends on the carbon dioxide content of the proteins which the maltine ultimately contains in suspension.

Brightness: malt should be bright. Turbidity in a malt may be due to deficiencies in filtration, microbiological contamination by bacteria or wild yeasts, presence of heavy proteins that were not removed during the fining process, degassing or oxygen contamination due to cracks in the capping, photochemical reactions due to overexposure to sunlight.

Quality specifications: maltine must specify the vitamin complex, consisting of vitamins B-1 and B-6, protein, calorific value, mineral content, Brix, etc.

1.9. Overview of the maltine production process

The maltine production process is similar to that of beer, with differences in the last stage of beer production, which is fermentation, since, as mentioned above, maltine does not require the fermentation stage. In this process, sorghum malt is first milled without being turned into flour, i.e. it is not pulverised, obtaining a granulated powder texture (Diaz, 2014). Then comes the mashing stage in which the malt is cooked at different temperatures. It is then filtered to extract all the liquid and washed to dissolve the sugar grains. The liquor is then cooked, where the remaining raw materials such as hops, caramel colour and raw sugar are added. A final stage of the process would be the cooling of the liquor and subsequent clarification, and then carbonation and finally cold storage (Gonzales & Ozuna, 2007).

1.10. Established technology for the production of maltine from sorghum

At present there is a well-defined technology for obtaining maltine from sorghum, which is referenced in (Ozuna, 2008) (Diaz, 2014) of the Central University "Marta Abreu" of Las Villas, which consists of the following stages: Malting (selection, soaking, germination y drying); Milling; Maceration; Filtration; Cooking o Boiling; Clarification; Cooling y Carbonation. The main characteristics of each of these stages are described below.

1.10.1. Malting

Malting is a process applied to cereal grains, in which the grains are germinated and dried rapidly after plant development. The malt is used to make beer, whisky and/or maltine. The malted grains develop the enzymes that are then needed to convert the starch in the grain into simpler sugars. Barley is the most commonly malted grain due to its high enzyme content. Other grains can be malted, although the resulting malt may not have sufficient enzyme content to convert its own starch content fully and efficiently (Owuama & Adeyemo, 2009).

Enzymes can be found in industrial form tai y сото are used today, but they can also be obtained due to changes occurring in a germinated grain. Hence, enzyme agents are used in the use of malt malt from germinated grains such as barley, wheat, sorghum, rye, etc.

The fundamental interest in malting is to obtain a grain that germinates easily and uniformly. Uniform o synchronised germination is very difficult if the kernels are not uniform in size, partly because larger kernels wet at a slower rate than smaller ones. Moreover, it is necessary that the kernels to be malted have not germinated before harvesting and that none of the kernels have died as a result of drying after harvesting in unsatisfactory circumstances. Malting requires that more than 98% of the kernels show the root pod. Malting also requires a low protein content, between 9 and 11.5% (Solange, Georgette, Gilbert, Marcellin, & Bassirou, 2014). The idea that lower starch content can also extend to the husk means that malting is looking for a grain with a low protein content y with poca husk. Finally, malting also aims to ensure that the malt performs well in its subsequent use in brewing, maltine o ethanol production, it must have a satisfactory enzyme endowment so that extraction is unproblematic. On the other hand, the wort must be easily separated from the spent grain y in relation to this, the grain must be poor in certain gums Hamada beta-glucans (Kayode, Hounhouigana, Nout, & Niehof, 2007).

In previous studies on sorghum malting (Carvajal, 2014) (Y. Diaz, 2014) (Gallardo et al., 2013), it has been shown that the quality of sorghum is responsible for its malting behaviour, which depends on several factors, e.g. whether it was harvested properly, whether it was fertilised with chemicals that influenced the sowing period, and whether it was previously dried to eliminate insects and fungi.

Malting corresponds to the first stages of germination and its objective is controlled germination through which enzymes (amylase, beta-glucanases and proteases) are produced to hydrolyse reserve materials. The malting process consists of the following stages (Gallardo et al., 2013) (Ozuna, 2008):

Selection

This stage is essential to ensure that the grain does not reach the germination process with a high microbial load, so moisture and protein content must be controlled.

Storage

Grain is more stable when dry and kept at low temperature. If it has been harvested with a moisture content of more than 15%, it is usually dried. The drying process must be carried out in such a way that the embryonic plant contained in each grain remains viable, therefore it is necessary to avoid using temperatures that are too high, while to increase drying it is necessary to use an increase in the air flow speed and gradual heating of the air. If the grain is wet, it is easily attacked by insects and fungi which cause deterioration. The metabolism of insects and fungi, when established, produces water and raises the temperature locally, which favours the spread of infection (Ortega, 2016).

Soaking

Typically, clean grain batches are dropped into a steeping tank partially filled with water at about 15 °C. The contents of the tank are intensively aerated by blowing air through the steeping water using perforated pipes o by suction, thus achieving the required 100% air. The contents of the tank are intensively aerated by blowing air through the steeping water using perforated pipes o by suction, thus achieving the required 100% air. Most soaking tanks are vertical tanks with a height of poca y piano bottom. This allows aerobic conditions to be created in the steeping water. The water content of the grains increases rapidly from the moment of immersion, but the rate of increase of the water content then decreases gradually. The speed of rewetting depends on the conditions in which the grain has grown, the variety and size of the grain, and the temperature of the water. It is also considerably influenced by any mechanical damage the grains may have suffered during soaking. Soaking is interrupted by draining after 12 to 24 hours. Each bean remains covered with a film of water through which oxygen from the surrounding air can dissolve. This condition is known as air resting. When the kernel has been soaked, the water penetrates through the husk y the fruit cover, y enters the kernel through the micropyle (Ortega, 2016).

The embryo takes up water quickly, while the endosperm hydrates more slowly. Any breakage of the husk ‾ the fruit coatings ‾ the seed facilitates the wetting of the endosperm ‾ the embryo and, of course, the escape of soluble substances from the endosperm. This is one of the factors contributing to losses during malting; another is the respiration of the embryo, which consumes nutrient reserves, releasing energy, carbon dioxide and water. Respiration increases significantly when the embryo is activated, which creates a demand for oxygen in the steeping water (A. Perez et al., 2010). In the absence of oxygen, the embryo can anaerobically metabolise the reserves, but in an energetically inefficient way, converting them to carbon dioxide and alcohol. As the concentration of alcohol increases, its toxicity increases, so it is necessary to change the soaking water from time to time.

In the case of sorghum, the soaking time is longer, depending on the physical conditions of the sorghum, the harvest and the initial moisture content (Gonzales & Ozuna, 2007).

Germination

Steeping is usually completed within two days. In modern malting techniques, the kernels give clear signs at the end of the malting process that they have begun to germinate, and are then transferred to the germination equipment. In most cases the moisture content is around 42 % and remains constant during the germination stage (Bernal, Perez, & Delgado, 2015).

Modern equipment allows germination in three or four days. The most common type of germinator is a box with a rectangular or circular base and a perforated false bottom. A bed of malt with a depth of 1 to 1.5 metres is placed on the false bottom. Through the bed, usually from bottom to top, a stream of water-saturated air is passed at about 15 °C, thus ensuring the availability of oxygen to the embryos, the removal of carbon dioxide and the maintenance of a constant temperature throughout the bed. To prevent rooting, a mechanical turner separates the germinating grains, which also helps to aerate and maintain a uniform temperature (Y. Diaz, 2014).

Sometimes a single vessel is used for soaking and germination, thus avoiding grain transfer, but often the soaking tanks are placed above the germination tanks. From the physiological point of view there is a continuity between soaking and germination. Embryo growth starts during soaking y como the immediately available nutrient reserves are limited, it is necessary to mobilise those of the endosperm (Nieblas et al., 2016). This alone would be insufficient to meet the needs of the rapidly growing embryo. These are supported by mobilisation of the aleurone layer, which produces enzymes from either complex precursors or amino acids. This mobilisation is triggered by a o further plant hormones called gibberellins which are secreted by the embryo o diffuse into the aleurone. Enzymatic degradation of the endosperm thus proceeds from the embryonic end of the kernel to the distal end of the kernel y from the outer to the inner layers. The physical weakening of the endosperm structure and the biochemical degradations are collectively known as 'deagglomeration'. Malted kernels can therefore be classified as under-disaggregated, disaggregated or over-disaggregated, depending on the extent of this enzymatic breakdown (Ramatoulaye et al., 2016).

Previous work (Chaviano, 2005) (L. Rodriguez, 2010) (Leyanis Rodriguez et al., 2015) (Nieblas et al., 2016) has found that the germination rate is highly dependent on the quality of the grain and the conditions in which it is found.

Drying

The aim of drying is to remove moisture, prevent further growth and modification, achieve a stable product that can be stored and transported, preserve enzymes, develop and stabilise flavour and colour properties, remove undesirable flavours, inhibit the formation of undesirable chemical compounds and dry the sprouts to allow their removal (Ramatoulaye et al., 2016).

The germination process is stopped by drying the malt kernels. At this stage, different options are available; you can obtain a malt that is not very disaggregated (LAGER malt), more disaggregated (ALE) for beer production or very disaggregated malt for use in distilleries or in vinegar production. You can also choose different kilning processes; prolonged dehydration y at low temperatures leads to light malt, with high enzyme content intact, while rapid dehydration y at high temperatures yields dark malts, deficient in enzyme activity (Y. Diaz, 2014). Numerous factors affect grain dehydration, including (Ortega, 2016):

- The volume of air passing through the grain bed.
- The depth of the bed.
- The weight of water to be removed from the grain bed.
- The temperature of the air used for dehydration.

Dehydration starts at temperatures of 50 - 60 °C, which initially heat the dryer and the grain bed. Later the upper layers start to dehydrate and the water content of the grains starts to gradually decrease from the bottom to the surface of the grain bed. In this stage of free dehydration, the water is removed from the grains without restriction, and for economic reasons the air flow is adjusted so that its relative humidity is 90-95% in the air at the outlet end (Ozuna, 2008). When approximately 60% of the water has been removed, subsequent dewatering is hampered by the nature of the residual water. At this breaking point, the temperature of the inlet air is raised and the flow rate is reduced. The thermal stability of the enzymes is now higher than when the malt contained 45 % water. When the water content reaches 12 %, all the water remaining in the grain is bound, so the inlet air temperature is raised to 65 - 75 °C and the flow rate is further reduced (Elgorashi, Elkhalifa, & Sulieman, 2016). Water removal is slow y for economic reasons much of the air is recirculated. Finally at a humidity of 5 - 8%, depending on the grain variety, the inlet air temperature is raised to 80 - 100 °C, until the required colour and humidity is reached. Typical Lager malts are kilned to a moisture content of 4.5 %, but Ale malts are dehydrated to a water content of 2-3 % (Carvajal, 2014).

Once the malting process has been carried out, the maltine production process is carried out, which consists of five stages o fundamental processes (Ozuna, 2008):

- Grain milling.
- Maceration
- Filtration.
- Cooking of the liquor.
- Cooling y clarification.

1.1.2. . Grinding

The grain is crushed in mills in order to break up the endosperm, but without crushing the glumillae.

1.1.3. . Maceration

The aim of this stage is to carry out the cleavage o hydrolysis of starch to maltose, which is more easily converted to the simpler form of sugar, glucose, by the action of enzymes (Nieblas et al., 2016).

At this stage, the ground beans are mixed with water, which has been previously heated. In this equipment the infusion maceration takes place, trying to obtain a homogeneous mixture, without lumps, from which the enzymatic extraction process begins. Temperature and time are two essential factors to be checked, since temperature steps and pauses at specific times ensure that all the important substances are transferred to the wort. In this process, the malt starches are converted into fermentable sugars, and the dissociation of proteins and organic phosphates, which have a significant influence on the acidity of the mash, is very important (Y. Diaz, 2014).

For all phases, the temperature is raised in a time interval of 15 minutes and maintained for 30 minutes. First the temperature is raised to 38 °C, known as the acidification temperature, to activate the organic phosphates, thus lowering the pH. Subsequently, the temperature is raised to 55-60 °C, the temperature at which proteolysis occurs to degrade the proteins into simpler proteins and amino acids. In the maceration process, the conversion of the amylaceous substances takes place. This process is known as saccharification, which is divided into two processes, first at 63 °C activating the enzyme p-amylase, which transforms starches into fermentable sugars, and then at 65 °C, where the enzyme a-amylase is activated, transforming starches into sugars y dextrins. The starch cleavage process involves the hydrolysis or saccharification of the linking points (bonds) of the glucose molecules producing a mixture of maltose and dextrins into fermentable sugars (Reyes, 2013).

1.1.3.1. . Parameters to be checked during maceration

Regardless of the name by which it is known, this technique is one of the oldest operations in the chemical industry. In the food and pharmaceutical industry it is used to recover important substances сото flavonoids сото carotenes o to remove undesirable substances сото contaminants o toxins. In all cases the extraction occurs сото result of the effect of the selectivity of the solvent with respect to the solute. From the industrial point of view it is necessary to evaluate some factors that influence the rate of extraction (Taylor & Dewar, 1994) (Ozuna, 2008) (Gallardo et al., 2013):

- **Temperature**

It is generally preferable to macerate at the highest possible temperature as this results in a higher solubility of the solute in the solvent and, consequently, it is possible to achieve higher final concentrations in the macerated decoction.

- **Degree of crushing**

The degree of grinding of the solid substance has a considerable influence on the rate of maceration, as grinding increases the contact surface between the phases y also reduces the path of the substance diffusing from the bottom of the pores to the surface of the solid material. However, as the degree of maceration increases, the energy consumption increases, and it should also be noted that maceration usually involves subsequent filtering, which is made more difficult as the particle size decreases.

- **Contact surface**

The accessible contact surface of the component to be extracted, for its interaction with the solvent, is shifted to the bottom of the pores of the solid material. This leads to a considerable decrease in the maceration rate when this parameter is limited by the internal diffusion velocity from the bottom of the grain or particle to its surface.

- **Choice of solvent**

Solvent selection is based on several factors including cost and toxicity, selectivity and dissolvability, as well as interfacial tension, viscosity, stability, and reactivity. The use of water as a solvent eliminates the difficulties of toxicity and treatment of other organic solvents. The chemistry of natural systems is based on water, so it is very common to use this cheap and not very dangerous solvent. Researchers in this area have found that reactions in water can optimise hydrophobic interactions, achieving high selectivities. The accelerating effect of water is due to several factors including the hydrophobic effect mentioned above as well as the hydrogen bonds between the water molecules and the reactants.

- **Itchy Liquid Agitation**

It allows to reduce the thickness of the diffusive boundary layer and to evenly distribute the solid particles in it, offers the possibility to considerably accelerate the maceration.

1.1.4. . Filtration

The extraction process, where the clear must and bran are separated by leaching, is achieved by recirculating the mash until a clear liquid is obtained, thus separating the husks and residues from the clean must. This first portion is called the first extract without the need to add water, followed by a second portion in which hot water is added to the bran to recover the maximum sugars contained in this substrate (Bernal et al., 2015). Once the extraction process is finished, the bran, which is rich in nutrients, is pumped to the receiving tank and marketed as animal feed.

1.1.5. . Cooking and boiling

Once filtered, the wort is boiled at 100 °C in order to inactivate the enzymes in the mash, sterilise the wort, solubilise and isomerise the bitter substances in the hops, especially the alpha-acids that form tannin-protein complexes that are soluble at high temperatures and insoluble at lower temperatures, and concentrate the wort, since additional water is added to the process by washing in the vat, which is removed by evaporation. During the boiling process, bitter hops, refined sugar, caramel colour and aromatic hops are added to the wort (Solange et al., 2014).

1.1.6. . Clarification

The boiled wort is sent to a settling tank equipped with a jacket in order to cool it y proceed to its clarification, thus favouring the sedimentation of coagulated materials (proteins, amino acids, p-glucans, starch, etc.) y the subsequent clarification of the maltine obtained (Maoura & Pourquie, 2009).

1.1.7. . Cooling y carbon dioxide saturation .

Once clarified, the maltine is sent to a vessel equipped with a coil, where it is cooled to temperatures of 0 - 2 °C, and then injected with carbon dioxide (CO2) in a process known as carbonation (Y. Diaz, 2014).

1.11. Overview of technological process design

In engineering, design can be defined as the creation of a system, component or process that satisfies a need. The successful operation of a future chemical plant or energy process will always depend on its design. The decision-making process involved in this process must be based on a correct knowledge of the raw materials to be used, the technologies that can be used, the price, the quality of the products, the pollution of the environment, as well as the control of a set of design and economic variables that affect the future competitiveness of the designed process (Towler & Sinnott, 2008). As previously defined, studies have been carried out at the Central University of Las Villas at various scales, mainly at laboratory (2 L) and pilot scale (1 hL), for the production of maltine from sorghum, specifically the UDG-110 bianco type, establishing an appropriate technological procedure for this type of chemical process (Ozuna, 2008) (Y. Diaz, 2014). However, no economic studies have been carried out on any scale with respect to obtaining maltine entirely from red sorghum CIAP R-132, so it is not known what economic indicators will be obtained by simulating a production plant of this type using professional process simulators.

1.12. Fundamental aspects of process simulation

In order to explain process simulation in chemical engineering, one must first turn to the history of simulation, then its definition and the areas it covers. In the first steps, process simulation was mainly based on analog circuits, using the phenomena of analogy. Indeed, systems theory shows that various physical principles have associated equivalent mathematical models o isometric ones. For example, certain electrical circuits, hydraulic circuits, processes of transfer of both matter сото energy and quantity of motion, are described by the same set of differential equations.

Consequently, it could be convenient to analyse (simulate analogically) the behaviour of a system (chemical process) by observing the evolution of the "equivalent" variables in an electrical circuit (whose model is equivalent -isomorphic- to the process studied), since they are easily measurable. Subsequently, with the massive use of the digital computer, and the revolution brought about by computer science in all fields of engineering, there has been a slow evolution from analogue to digital simulation, with the former having practically disappeared in many applications (Valle, 2013) (E. J. Perez, 2016).

As is well known, the computer has been used for engineering calculations for only a few decades. In particular, a computer is characterised simply by the fact that it performs calculations quickly after being programmed. It stores, manipulates and gives rapid access to enormous amounts of information, while allowing symbolic manipulations to be performed. Regardless of the way in which this is achieved, what is important in our case is to understand the consequences; that is, the implications of this phenomenon in the field of engineering. In 1974, the first chemical process simulator (FORTRAN) appeared. Since then, a succession of developments has led to the existence of several efficient commercial simulators, such as SPEED UP, ASPEN PLUS, SuperPro Designer, HYSYM, HYSYS, ChemCAD, etc. (Scenna, 1999).

1.13. Classification of simulation methods

A simulation task can be considered to be one in which certain input values are inserted into the simulator o simulation program to obtain certain results o output values, thus estimating the actual behaviour of the system under those conditions. Simulation tools can be classified according to various criteria, for example, according to the type of process (batch o continuous), whether it involves time (stationary o dynamic -including batch equipment-), whether it handles stochastic o deterministic variables, quantitative o qualitative variables, etc. In the following, the characteristics of the different types of simulation tools generally used will be briefly outlined (Law, 2011).

1.13.1. Stationary simulations

The steady state simulation involves solving the balances of a system without involving the time variable. If the model wishes to study the variations of the variables of interest with spatial coordinates, it will be a distributed parameter model. An example could be the radial variation of the composition in a distillation column plate, the variation of the properties with length and radius in a tubular reactor, etc. Generally, in commercial simulators, concentrated parameter models are used.

1.13.2. Simulaciondinamica

On the other hand, dynamic simulation considers the balances in their dependence on time, either to represent the behaviour of batch equipment or to analyse the evolution that occurs in the transition between two stationary states for a piece of equipment or a complete plant. In this case, the mathematical model will be constituted by a system of ordinary differential equations whose differential variable is time, in the case of models with concentrated parameters. Otherwise, a system of partial derivative differential equations must be solved, covering both the spatial and temporal coordinates (distributed parameters).

1.13.3. Qualitative simulation

The main purpose of qualitative simulation is the study of causal relationships and qualitative time trends of a system, as well as the propagation of disturbances through a given process. We call qualitative values of a variable, as opposed to numerical (quantitative) values, its sign, either absolute or relative to a given reference value. Therefore, in general, we work with such values сото (+, -, 0).

There are several fields of application of qualitative simulation, e.g. trend analysis, monitoring and fault diagnosis, alarm analysis and interpretation, statistical process control, etc.

1.13.4. Quantitative simulation

Quantitative simulation, on the other hand, describes the behaviour of a process numerically, using a mathematical model of the process. This is done by solving the balances of matter, energy and quantity of movement, together with the constraint equations that impose functional and operational aspects of the system. The quantitative simulation covers mainly steady-state and dynamic-state simulation.

Other classifications

From the point of view of the phenomena or systems being studied, simulation can also be classified as deterministic or stochastic.

A deterministic model is one in which the equations depend on parameters y variables known with certainty, and there is no uncertainty or probability laws associated with them. In contrast, in a stochastic model, certain variables will be subject to uncertainty, which may be expressed by probability distribution functions (Martinez, 2012).

1.13.5. Types of simulators

Process simulators can be divided into the following types, depending on the philosophy under which the mathematical model representing the process to be simulated is proposed (Chung, 2008):

- Global or equation-oriented (mathematical) simulators
- Modular simulators
- Hybrid simulators o sequential-simultaneous modular

Global simulators

These are those where the mathematical model that represents the process is proposed by means of the elaboration of a large system of algebraic equations that includes the whole set of o plant to be simulated. In this way, the problem is translated into solving a large system of algebraic equations, usually highly non-linear. Among the main features of these simulators are the following:

- Each piece of equipment is represented by the equations that model it. The model is the integration of all subsystems.

- The distinction between process variables and operational parameters disappears, thus simplifying design problems.
- Simultaneous solution of the resulting system of (non-linear) algebraic equations.
- Resolution of the Equations entered by the user that make up the model of the process under study.
- It is required to define the computational algorithm y use numerical methods y advanced ordering techniques y decomposition of equations to find the solution of the mathematical problem posed.
- Higher convergence speed.
- More difficult to use by "non-specialists".

Within this group of simulators are: MATLAB, MATHEMATICA, GAMS, XPRES, etc. ((Scenna, 1999).

Modular simulators

These are based on independent simulation modules that follow approximately the same philosophy as the unitary operations, i.e. each equipment: pump, valve, exchangers, etc.; are modelled through specific models for them and, in addition, the direction of the information coincides with the "physical flow" in the plant. Conceptually, under this philosophy, for each simulation module (equipment) a mathematical model must be considered. The main features are shown below (Scenna, 1999):

- Efficiently solved individual models.
- Easily understood by "non-simulation" engineers.
- The information entered by the user (related to current equipment) is easily checked and interpreted.
- Design problems (selection of parameters) are more difficult to solve.
- Difficulty is increased when an optimisation problem is posed (black boxes work).
- Not very versatile, but very flexible, very reliable and quite robust.

1.14. Advantages and disadvantages of process simulation

According to (Gonzalez, 2013) simulation has сото main advantages and disadvantages:

Advantages

- It is a relatively efficient and flexible process.
- It can be used to analyse y synthesise a complex y large real situation, but it cannot be used to solve a conventional quantitative analysis model.
- In some cases simulation is the only method available.
- Simulation models are structured to solve transcendent problems in general.

- The directives require to know сото is advanced y which options are attractive; the directive with the help of the computer can obtain various decision options.
- Simulation does not interfere with real-world systems.
- Simulation makes it possible to study the interactive effects of individual components o variables to determine the most important ones.
- Simulation allows the inclusion of real-world complications.

Disadvantages

- A good simulation model can be quite costly; often the process of developing a model is a long and complicated one.
- Simulation does not generate optimal solutions to quantitative analysis problems, in techniques сото economic order quantity, linear programming. By trial and error different results are produced in repeated runs on the computer.
- The directives generate all the conditions and constraints to analyse the solutions. The simulation model does not produce answers by itself.
- Each simulation model is unique. Solutions and inferences are usually not transferable to other problems.
- There will always be variables left out y those variables (if there is bad luck) can completely change the results in real life that the simulation does not foresee... in engineering you "minimise risks, not avoid them".

1.14.1. Features of the main chemical process simulators

ASPEN PLUS

Developed by Aspen Technology, Inc., ASPEN PLUS is a stationary simulator, as well as a modular sequential simulator (in the latest versions it allows equation-oriented strategy). Oriented to the process industry, chemical and petrochemical, it models and simulates any type of process for which there is a continuous flow of materials and energy from one process unit to another.

This simulator allows:

- Regression of experimental data.
- Preliminary design of flow diagrams using simplified equipment models.
- Perform rigorous material and energy balances using detailed equipment models.
- Dimensioning key pieces of equipment.
- On-line optimisation of complete process units or plants.CHEMCAD.
- Developed by Chemstations, CHEMCAD has evolved with the industry. We believe in the value that chemical engineers bring to the modern world and are dedicated to providing tools that help advance the field of process engineering. It consists of:

- CC-STEADY STATE: Ideal for: Users who wish to design processes, o existing, steady-state type processes.
- CC-DYNAMICS: Ideal for: Users who want to design processes o dynamic type.
- CC-THERM: Ideal for: design of a heat exchanger (a single unit at a time), y those who want to existing type exchangers in a new service o carry out.
- CC-SAFETY NET: Combines the latest in two-phase relief device calculation, rigorous pressure drop calculation, rigorous physical properties calculation, and rigorous phase equilibrium calculation to provide fast and accurate responses. Ideal for: Users who need to design networks o type of piping o safety relief devices y systems.
- CC-FLASH: Physical properties and phase equilibrium calculation software that is a subset of the CHEMCAD suite (all products in the CHEMCAD suite include CC-FLASH capabilities).
- CC-BATCH: Ideal for: Users who need property y phase equilibrium data, as well сото users who need property prediction y regression.

HYSYS

HYSYS is a software, used to simulate steady-state and dynamic processes, e.g. chemical, pharmaceutical, food and other processes.

It has tools that allow us to estimate physical properties, matter and energy balance, liquid-vapour equilibria and the simulation of many chemical engineering equipment. This simulator in its latest versions allows to use o create operator models. The design parameters сото number of tubes of a heat exchanger, shell diameter y number of plates of a distillation column cannot be calculated by HYSYS, it is a tool that provides a simulation of a system described above. HYSYS can be used as a design tool, testing various system configurations to optimise the system.

SuperPro Designer

Developed by Intelligen, Inc. is one of the most comprehensive and well known process design and simulation packages, being a very versatile simulator that can meet the needs of engineers in a wide variety of industries, such as Biotechnology, Pharmaceutical, Chemical, Food, Mining, Wastewater Treatment, Environmental Control, etc. It combines different models of unit operations that allows the user to simultaneously design and evaluate processes. It is a very useful tool for the scientific and engineering development of products and processes. It enables efficient development, evaluation and optimisation of technologies from an environmental point of view. It has an interface developed in a user-friendly Windows environment and the results can be exported to compatible Excel, Lotus, etc. spreadsheets

(Valle, 2013).

MATLAB

Developed by MathWorks, MATLAB (short for MATrix LABoratory) is a mathematical software tool that provides an integrated development environment (IDE) with a proprietary programming language (M language) and species service. It is available for Unix, Windows, Mac OS X and GNU/Linux platforms. Its basic features include: matrix manipulation, data representation y functions, implementation of algorithms, creation of user interfaces (GUIs) y communication with programs in other languages y with other hardware devices. The MATLAB package has two additional tools that expand its capabilities, namely Simulink (multi-domain simulation platform) and GUIDE (GUI editor). In addition, MATLAB's capabilities can be extended with toolboxes; y Simulink's capabilities can be extended with blocksets.

It is widely used in universities and research and development centres. In recent years it has increased the number of features, such as the ability to directly program digital signal processors or create VHDL code (Jalon, 2012).

Toolboxes y block packages MATLAB functionalities are grouped into more than 35 toolboxes y block packages (for Simulink).

1.15. The SuperPro Designer process simulator.

Simulation software makes it possible to simulate various industrial processes and operations on a computer. In many fields, the chemical and petroleum industries have developed specific simulators. Among the many existing simulators, we can say that the ***SuperPro Designer^*** is one of the most relevant at present, this has been used for the simulation of various industrial processes having a great effectiveness (Intelligen, 2014).

The simulator is useful for improving new designs and modifying existing operations to ensure that equipment is working within specifications. A simulator does not optimise by answering the "what if" question, which allows us to improve a design, or make a plant run better, or improve the quality of the products manufactured by allowing us to analyse alternatives that we would study without the simulator. Computer simulation solves this system of algebraic and/or differential equations.

1.16. Computer-aided chemical plant simulation

The simulation of a chemical plant involves the creation of a process model, where a model is understood as a description of the behaviour of a real process, capable of predicting the output (responses) as a function of the inputs. Process simulation for the evaluation of alternatives and loss mitigation in a given process will play a very important role for the final optimisation model, as it can reveal on a large scale whether a future process can give successful results or not (Ruiz-Mercado & Cabezas, 2016).

These tools can be employed at all stages of process development, from conceptual process design to operation and subsequent plant optimisation (Auli, Sakinah, Bakri, Kamarudin, & Norazian, 2013).

Process simulators offer the opportunity to shorten the time required for process development. They allow the comparison of process alternatives on a consistent basis, so that a large number of process ideas can be synthesised and analysed interactively in a short period of time.

Despite some anticipated differences between process simulation and real-time plant operation, process simulators are commonly employed to provide reliable information about process operation, taking into account their extensive database of chemical components, thermodynamic packages and advanced computational methods (Auli et al., 2013).

1.17. Technical-economic analysis. NPV, IRR and PRI calculations.

The fundamental concepts and mathematical operations necessary for basic economic analysis are basic to process design. Profitability calculations are combined with operations research techniques, so that each alternative being evaluated in the design can be obtained with a justifiable amount of effort. Among the most commonly used techniques are: the *payback* period, which gives a quick indication of when the investment is attractive; the return on investment (ROI), which is widely used in preliminary calculations in order to get an idea of the profitability of a project. In addition, dynamic calculations are used, which are based on discounted cash flow and have broad advantages over ROI because they consider the time value of money by assuming that costs precede revenues (Baca, 2004).

The company generally adds the cost of risk associated with each investment project according to its characteristics, measured in terms of entrepreneurial criteria, market situation, historical experience and projection. Currently, interest rates between 10 and 15% are recommended for the chemical industry (Dominguez, 1996).

The dynamic methods for investment analysis are the following: Net Present Value (NPV), Internal Rate of Return (IRR) and Investment Recovery Period (IRP). These methods complement each other and are not mutually exclusive, hence the desirability of applying them together in practice (Perry & Green, 2008).

1.18. Key process indicators for sensitivity analysis

Sensitivity analysis is a tool used to estimate the contribution of input factors to output variables. In this case, the aim is to determine the possible variation of the initial outlay, the cash flows and the discount rate in order to make the investment worthwhile. This analysis is carried out according to the NPV y according to the IRR. The sensitivity analysis to determine the profitability of an investment, it can be said that it is a financial analysis that is essential to measure the performance of the business and to identify the problems that may arise (Martin, 2016).

A financial study evaluates the project taking into consideration a large number of assumptions for each of the variables involved in the analysis, which together form the "most likely" scenario o the base case. The conclusions obtained are therefore always valid in the light of these assumptions, which gives them a high degree of rigidity. In other words, the outcome of the financial study is static, in that it does not in itself allow comparative conclusions to be drawn, other than the acceptability of the project under the criteria adopted. With the aim of obtaining as much background information as possible in order to formulate an investment decision on a more solid basis, this section will develop a sensitivity analysis, which will allow us to measure the changes in the results of the evaluation in the face of changes in the variables that make up the flow of funds. The aim is to deepen the study by giving it the character of comparative statics (Moran, 2015).

First, certain methodological considerations to be taken into account for the analysis will be summarised. Secondly, the structure of the flow of funds will be outlined in order to identify the most relevant variables to be studied. Thirdly, the unidimensional sensitisation of these variables will be developed. Fourth, a two-dimensional sensitisation model will be developed to measure the effects of simultaneous changes in two variables (Towler & Sinnott, 2008).

In a simple sensitivity analysis, each parameter is individually varied, and the output is a qualitative understanding of which parameters have the greatest impact on the viability of the project. In a more formal risk analysis, statistical methods are used to examine the effect of variation in all parameters simultaneously and quantitatively determine the range of variability in economic criteria, allowing the plan engineer to estimate the degree of reliability of the chosen economic criteria (Perry & Green, 2008).

1.19. Economic indicators for the evaluation of investment proposals (NPV, IRR, IRR, PRI)

Economic engineering in its most general form can be defined сото a collection of mathematical techniques that simplify economic comparisons. With the help of these techniques, a comprehensible and rational procedure can be developed to evaluate the economic aspects of different proposed methods. As an evaluation criterion for comparing such methods in economic engineering, money is used as the main criterion.

Net present value, also known as net present value (*NPV)*, is a procedure for calculating the present value of a given number of future cash flows arising from an investment. The methodology consists of discounting all the future cash flows of the project to the present time (i.e. discounting at a rate). The initial investment is subtracted from this value, so that the value obtained is the net present value of the project (Baca, 2004).

The present value method is one of the most widely used economic criteria in the evaluation of investment projects. It consists of determining the equivalence in time 0 of the future cash flows generated by a project and comparing this equivalence with the initial outlay. When this equivalence is greater than the initial outlay, then it is recommended that the project be accepted.

The formula that allows us to calculate the Net Present Value is (Baca, 2004):

$$\mathrm{VAN} = \sum_{t=1}^{n} \frac{V_t}{(1+k)^t} - I_0$$

Where:

- Vt - cash flows in each period t.
- *Io* - value of the initial outlay of the investment.
- *n* - number of periods considered.
- *κ* - type of interest.

If the project has no risk, the fixed income rate will be used as a reference, so that the NPV will be used to estimate whether the investment is better than investing in something safe, without specific risk. In other cases, the opportunity cost is used. When the NPV takes a value equal to 0, *κ* is called IRR (*Internal Rate of Return*). The IRR is the profitability that the project is providing us (Martin, 2016).

The **internal rate of return** o internal rate of return (IRR) of an investment is the geometric average of the expected future returns on that investment, y which incidentally implies the assumption of an opportunity to "reinvest". In simple terms, various authors conceptualise it as сото the discount rate at which the net present value o net present value (NPV o NPV) equals zero.

IRR can be used сото indicator of the profitability of a project: the higher the IRR, the higher the profitability, thus, it is used сото one of the criteria for deciding on the acceptance or rejection of an investment project. For this purpose, the IRR is compared with a minimum rate o cut-off rate, the opportunity cost of the investment (if the investment is risk-free, the opportunity cost used to compare the IRR will be the risk-free rate of return). If the project's rate of return expressed by the IRR exceeds the cut-off rate, the investment is accepted; otherwise, it is rejected (Martin, 2016).

The Internal Rate of Return IRR is the discount rate that makes the NPV equal to zero (Baca, 2004):

$$VAN = \sum_{t=1}^{n} \frac{F_t}{(1+TIR)^t} - I = 0$$

Where:

- *Ft - Cash* flow in period *t.*
- *n* - Number of periods.
- *I* - Value of the initial investment.

The Schneider approximation uses the binomial theorem to obtain a first order formula:

$$(1+TIR)^{-n} \approx 1 - n * TIR$$

$$\bar{I} = F_1 * (1 - TIR) + F_2 * (1 - 2 * TIR) + ... + F_n * (1 - n * TIR)$$

$$TIR = \frac{-I + \sum_{i=1}^{n} F_i}{\sum_{i=1}^{n} i * F_i}$$

However, the calculation obtained may be far from the actual IRR.

The **payback period (PRI)** is the time (days, months, years) after which the investment made is fully recovered. It is an important measure of the economic recoverability and profitability of any project, i.e. the lower its value, the faster it will generate profits (Grech, 2002).

1.20. Effects on the environment

Cuba has entered into a profound process of developing its policy and strategy for the protection of the environment, which contributes to the elimination of risks, to creating new market opportunities and technological innovation, to increasing the motivation of workers and to strengthening business-society relations (Gomez, 1999).

1.21. Coeliac disease

Coeliac disease is a lifelong autoimmune disease, in which sorghum is not a drug, but part of the diet of these patients from a gluten-free product o protein that is harmful to these patients.

It is a disorder that occurs in genetically predisposed individuals of all ages, starting in infancy. Symptoms include chronic diarrhoea, retarded growth and/or childhood development, dyspnoea, skin rashes, weight loss, weight loss, changes in character, vomiting and bloated belly. Although these symptoms may be absent and appear from time to time, they can also appear in almost every organ and system of the body. The disease affects an estimated 1% of the population of Indo-European ethnicities, although it is thought to be a considerably under-diagnosed disease. As a result of early screening, an increasing number of asymptomatic diagnoses are being observed. The only effective treatment is a lifelong switch to a gluten-free diet, which allows regeneration of the intestinal villi (Pino, 2017).

Cuba has some 250,000 adults suffering from the disease, making it a promising alternative for the food of these patients, as well as helping to substitute imports, given the advantages of its cultivation (Ortega, 2016).

CHAPTER 2

Materials and methods

2.1. Description of the maltine production process from red sorghum CIAP R-132.

Annex 1 shows the block diagram containing the main operations involved in the production process of maltine from red sorghum CIAP R-132. This process consists of six fundamental stages, which are:

2.1.1. Moliendadelgrano.

2.1.2. Maceration.

2.1.3. Filtration y washing.

2.1.4. Cooking.

2.1.5. Primary cooling y clarification.

2.1.6. Secondary cooling y carbonation.

2.1.7. Grain milling

22.50 kg of CIAP R-132 red sorghum grains are ground in a disc mill to the consistency of granulated flour.

2.1.8. Maceration

The maceration consists of 4 temperature pauses (Annex 3), where first the water contained in the macerator is heated to a temperature of 38 °C, and then the previously ground sorghum is added. The sorghum-water mixture is kept at this temperature for 40 minutes. After this time, the mashing temperature is increased to 52 °C, and the mashing temperature is maintained at this value for 40 minutes. During this pause, the enzyme liquor Fungamyl 800L is added at a concentration of 0.1 g/L. The process temperature is then increased to 63 °C and maintained at this temperature for a further 40 minutes. The temperature is then increased to 71 °C and the macerate is kept at this value for 60 minutes. At the end of this time, the temperature of the mixture is increased to 78 °C and kept at this value for 10 minutes in order to stop the enzyme activity. The enzymatic starch hydrolysis reaction is described by the following equation (Reyna et al., 2004):

$$2(C_6 H_{10} O_5)_4 + nH_2 O \wedge nC_{12} H_{22} O''$$

With the application of these temperature pauses, the starch contained in the sorghum is correctly saccharified into fermentable sugars. The whole mashing process is carried out at a

derpm stirring speed , and the mashing that takes place during the mashing process

is

is carried out in this process is the infusion type. At the end of the mashing process, the must obtained is tested for iodine (Annex 4).

2.1.9. Filtration y washing

The mixture obtained at the end of the mashing process is filtered through a sieve to obtain the filtered must (liquid) and the wet sorghum grain (bran). The bran is washed with hot water at a temperature of 70 °C to dissolve any remaining sugars in the solid, while the wort is pumped back to the masher for boiling.

2.1.10. Preparation of sucrose

In a separate vessel, 10,50 kg of raw sugar is diluted in 15 L of water, and the resulting mixture (sucrose) is sterilised at 121 °C for 20 minutes. After sterilisation, the sucrose is cooled.

2.1.11. Cooking

The wort in the mash tun is heated to 100 °C, where 120 g of bitter hops are added 10 minutes into the boiling process, while 13.95 kg of caramel and 30 g of aroma hops are added 15 minutes before the end of the boiling process. All the previously sterilised sugar is added at the beginning of the boiling process. The wort is kept at 100 °C for 60 minutes while the mixture is stirred at a stirring speed of 150 rpm.

2.1.12. Primary cooling y clarification

At the end of the boiling stage, the hot wort is pumped into a whirlpool settler, where it is cooled to a temperature of ± 30 °C by circulating cooling water through the jacket of the apparatus. This cooling causes coagulable particles, impurities and other settleable substances to settle to the bottom of the apparatus, while the supernatant produces the clarified maltine.

2.1.13. Secondary cooling y clarification

Once the clarified liquid reaches 30 °C inside the settler, it is pumped into the reactor, where its temperature is reduced to 1-2 °C by circulating alcoholic water inside the reactor coil. When this temperature is reached, the cold maltine is carbonated by injecting carbon dioxide (CO2) gas until it is completely saturated, resulting in cold carbonated maltine ready for consumption.

2.2. Simulation of the production process in the SuperPro Designer® Simulator

The maltine production process from red sorghum CIAP R-132 was simulated using the simulator SuperPro Designer® v. 8.5, applying the tools of mass and energy balance, equipment design and economic calculations contained in this software (see Annex 2). This allowed to obtain a number of technical and economic parameters of the process under study such as NPV, IRR, IRR, IRP, fixed costs, operating costs, unit cost of production, gross and net margin, % return on investment, among others.

During the simulation it was assumed that the plant has a construction period of 12 months, with 2 months for start-up and commissioning. The production capacity of the plant is taken as 150 L/batch, as this is the average production capacity of the Pilot Plant of the University of Camaguey. A life time of the project of 15 years was taken into account, where the plant produces at 100% capacity during its entire life time. The NPV value was determined considering an interest rate of 11 %, and 32 % income tax was taken into account. The cost of validation and start-up was assumed to be 15% of the Direct Fixed Capital (DFC), the costs related to Quality Assurance and Quality Control were assumed to be 15% of the Total Labour Cost, and an average wage of $ 3.00/hour was applied for the operators working in the plant, and $ 5.00/hour for the supervisors and management personnel.

It was also assumed that there is no product rejection due to non-compliance with established quality standards, that about $3,000 is spent annually on validation of the production process, and that the plant's waste treatment costs constitute 20% of the total operating costs.

The plant consumes cooling water, electricity and steam, and operates 11 months a year, using one month for maintenance and repair of equipment and auxiliary systems. Table 2 shows the cost of purchasing the main raw materials, the materials used, and the selling price of the products obtained in the production process.

Table 2. Acquisition cost of the main raw materials and products obtained in the production process, which were used in the simulation.

Compound	Price	Unit
Sorghum redCIAP R-132	14,05	$/kg
Water	0,10	$/m^3
Bitter hops	51,228	$/kg
Aromatic hops	65,31	$/kg
Raw sugar	356,15	$/Ton
Afrecho	60	$/Ton
Bottle 350 mL	0,306	$/U
Covers	0,007	$/U
Maltina in 350 mL bottles	6	$/bottle

Annex 5 shows the acquisition costs of each piece of equipment used in the production process, according to the literature consulted (Baca, 2010; Perry & Green, 2008; Peters & Timmerhaus, 1991; Sinnott, 2005; Towler & Sinnott, 2008; www.matche.com, 2019). The final product obtained (maltine) will be packaged in 350 mL glass bottles.

Annex 6 shows the percentage chemical composition of sorghum, hops and caramel colour used in the simulation (Diaz, 2014) (Perez, 2016).

Finally, a plant production capacity of 150 litres of beer per batch is chosen, as this is the average production capacity of the Pilot Plant of the University of Camaguey.

2.3. Assessment of the sensitivity of selected indicators

Once the maltine production process was simulated in the SuperPro Designer® simulator, we proceeded to develop a design of experiments of the Response Surface type using the statistical software *Statgraphics Centurion XVI®,* with the aim of evaluating the statistical influence of 3 initial input variables o, which are (1) Maltine production capacity per batch of the plant; (2) Cost of acquisition of sorghum y (3) Sale price of the bottle of maltine, on three important output parameters of the process: NPV, IRR y PRI. Therefore a sensitivity study containing 3 inputs and 3 outputs will be carried out.

In order to make the previously described design of experiments, the values used during the simulation for the 3 initial variables considered will be increased or decreased by 20 %, in order to take into account possible variations or oscillations of the values that these variables may present in the future, with the aim of evaluating the potential influence that these variations may have on the results to be obtained from the NPV, IRR and PRI, as well as to select the most feasible run (or scenario) from the economic point of view.

A randomised *Response Surface* type design of experiments was developed by applying the *"Box-Behnken Design"* option contained in the *Statgraphics®* statistical package, from which a total of 30 experimental runs were initially obtained (see Annex 7), which were subsequently optimised by applying the *"D-Optimality"* tool contained in the statistical software itself, with the aim of selecting those runs that have the greatest statistical influence on the 3 output variables taken into account, thus сото also reduce the length of the sensitivity study, finally reaching 11 experimental runs (see Annex 8). The selected optimal experimental runs are shown in red (Annex 8), while Annex 9 shows the values that each of the 3 input variables should present taking into account the range of ± 20%, y the values that these 3 parameters should present within the optimised design of experiments containing the 11 experimental runs.

2.6. Determination of the statistical correlation between the input variables and the indicators NPV, HR and PRI.

Once the results of the sensitivity study were obtained, we proceeded to analyse the statistical correlation between the three input variables considered and the three output parameters evaluated, in order to obtain equations that quantitatively describe the statistical relationship between each of the input variables and the output variables. This was carried out using the *"Multiple Regression"* option of the statistical package *Statgraphics CenturionXVI®.*

Also determine the run or experiment with the most positive economic result in relation to the NPV, IRR and PRI results obtained, as well as the one with the most negative result.

CHAPTER 3

Analysis of the results

3.1. Main technical-economic results obtained during the simulation of the maltine production process at pilot scale.

Table 3 shows the values of the main technical-economic indicators obtained during the simulation of the maltine production process in the SuperPro Designer® simulator. Annex 10 shows the rest of the technical-economic results obtained.

Table 3. Main technical-economic indicators obtained during the simulation of the maltine production process in the SuperPro Designer® simulator.

Indicator	*Value*
Total Capital Investment [$].	613 000
Operating Cost [$/year].	1 507 000
Total annual earnings [$/year].	2 422 000
Unit Cost of Production [$/bottle].	3,73
Return on Investment [%] Return on Investment [%] Return on Investment [%] Return on Investment [%] Return on Investment [%] Return on Investment	98,02
Return on Investment Period (ROI) [years].	1,02
Internal Rate of Return (IRR) [%].	63,20
Net Present Value (NPV) [$].	3 661 000

Taking into account the results shown in Table 3, the project can be considered economically profitable and reliable from an investor's point of view (Baca, 2005; Baca, 2010; Sinnott, 2005; Towler & Sinnott, 2008), since the NPV has a positive sign ($ 3 661 000), the IRR is higher than 25% (63.20%) and the IRR is lower than 5 years (1.02), which is synonymous with profitability and profitability.

Annex 11 shows the breakdown of the main raw materials and materials consumed per year, and their percentage influence on the total cost of production. Annex 14 shows the Gantt Chart obtained using the SuperPro Designer® simulator.

In (Ozuna, 2008) it was obtained that a maltine production plant with a capacity of 159 Ton/a is not feasible from the economic point of view, since it has a NPV of $-1 349 729.75, without obtaining IRR y PRI values.

In (Diaz, 2014) an economic analysis was carried out for the production of 225 ton/year of maltine from UDG-110 white sorghum, where a net present value of $ 318 591.73, an interest rate of 37% and a payback period of 3.4 years was obtained.

3.2. Evaluation of the sensitivity study carried out with respect to the initial variables selected.

Annex 11 presents the results of the sensitivity study carried out taking into account the 11 runs evaluated. According to the results presented in the same, runs # 3 and # 9 are the ones that present the most positive economic scenario with respect to the NPV values ($ 7

997 968), IRR (103.36 %) and IRR (0.51 years). This is due to the fact that in these runs the highest value of production capacity per batch (180 L/batch), the lowest sorghum purchase cost ($ 11.24/kg) and the highest maltine selling price ($ 7.2/bottle) are obtained. That is, the highest possible production capacity is produced with the lowest acquisition cost of the main raw material consumed (sorghum), and the final product obtained (bottle of maltine) is sold at the highest possible price, which has a positive impact on the overall profitability of the production process.

On the other hand, run # 2 presents the worst economic scenario of all, with a negative NPV value (- $ 13 311), an IRR of 6.64% and a PRI of 7.11 years. This is due to the fact that this run has the lowest production capacity per batch (120 L/batch), the highest purchase cost of sorghum ($ 16.86/kg) and the lowest selling price of the maltine bottle ($ 4.8/bottle).

3.3. Results obtained with respect to the statistical correlation between the 3 input variables and the indicators NPV, IRR and PRI.

By evaluating the statistical correlation between the three initial input variables considered (maltine production capacity per batch; cost of purchasing red sorghum; and selling price of the maltine bottle) and three important economic indicators of the process (NPV, IRR and PRI), the following equations were obtained:

- **Net Present Value**

NPV = -1.43956E7 + 57633,9*Production capacity - 99797,7*Sorghum cost + 1,77474E6*Maltine sales

- **Internal Rate of Return**

IRR = -151.032 + 0.659782*Production capacity - 1.14981*Sorghum cost + 20.7559*Maltine sales

- **Investment Payback Period**

PRI = 12,5009 - 0,0356912*Production capacity + 0,0874503*Sorghum cost - 1,05221*Maltine sale

CHAPTER 4

Conclusions

1. The maltine production process from red sorghum CIAP R-132 at pilot scale (150 L/batch) can be considered economically profitable considering the NPV, IRR and PRI values obtained, which were $ 3 661 000, 63.20 % and 1.02 years, respectively.
2. The runs that presented the best economic scenario during the sensitivity study were # 3 and # 9 with a NPV = $ 7 997 968, an IRR = 103.36 % and a PRI = 0.51 years, while the worst economic result was # 2 with a NPV =-$13311, an IRR = 6.64 % and PRI = 7.11 years.
3. Statistical correlations were obtained to quantitatively relate the NPV, IRR and IRR indicators to three initial parameters of the production process: 1) Maltine production capacity per batch; 2) Red sorghum purchase cost; and 3) Maltine bottle selling price.

Recommendations

1. To carry out simulations related to maltine production on a pilot scale using other sorghum varieties.
2. To carry out technical-economic studies to determine the feasibility and profitability limits of the maltine production process on a pilot scale.
3. Producing maltine from CIAP R-132 red sorghum at pilot plant scale.
4. To carry out industrial scale simulations of the maltine production process using red sorghum CIAP R-132 (1000 L/batch).

Bibliographical References

Agu, R. C., & Palmer, G. H. (1997). The effect of temperature on the modification of sorghum and barley during malting. *Process Biochem,* 32, 501-507.

Agu, R. C., & Palmer, G. H. (1998). A reassessment ofsorghum for larger-beer brewing. *Bioresour. Technol.,* 66, 253-261.

Almodares, A., & Hadi, M. R. (2009). Production of bioethanol from sweet sorghum: A review. *African Journal ofAgricultural Research*, 4(9), 772 - 780.

Auli, N. A., Sakinah, M., Bakri, A. M. M. M. A., Kamarudin, H., & Norazian, M. N. (2013). Simulation Of Xylitol Production: A Review. *Australian Journal of Basic and Applied Sciences*, 7(5), 366-372.

Baca, G. (2004). *Ingenieria Economica (*8th ed.). Bogota, Colombia: Fondo Educativo Panamericano.

Bernal, L. G., Perez, G. I., & Delgado, R. (2015). *Transformation and innovation of grains: sorghum for craft beer brewing* Paper presented at the 20th National Meeting on Regional Development in Mexico, Cuernavaca, Morelos.

Carvajal, N. (2014). *Perfeccionamiento delproceso de producción de cerveza a partirde malta de sorgo.* Diploma thesis, Department of Chemical Engineering, Faculty of Chemistry-Pharmacy, Universidad Central "Marta Abreu" de Las Villas, Santa Clara, Cuba.

Chaviano, M. (2005). Sorghum: Contribution to the sustainable and ecological development of popular rice production. *Agricultura Organica,* 1,8-11.

Chung, C. A. (2008). *Simulation Modeling Handbook; A Practical Approach* (3rd ed.). Boca Raton: CRC Press.

Dewar, J., Taylor, J. R. N., & Berjak, P. (1997). Determination of improved steeping conditionsforsorghum malting. *J. CerealSci.,* 26, 129-136.

Diaz, Y. (2014). *Improvement of the sorghum malting process for the production of maltines for celiac patients.* Diploma thesis, Universidad Central "Marta Abreu" de Las Villas Villa Clara, Cuba.

Doggett, H. (1998). *Sorghum* (2nd ed.). London, UK: Longman Scientific and Technical.

Dominguez, E. R. (1996). *Analysis of investment alternatives in the chemical industry considering equipment reliability.* Diploma thesis, Universidad Central "Marta Abreu" de Las Villas, Santa Clara, Cuba.

Douglas, J. M. (1988). *Conceptual Design of Chemical Process.* New York, USA: McGraw-Hill.

Elgorashi, A. G. M" Elkhalifa, E. A., & Sulieman, A. m. E. (2016). The Effect of Malting Conditions on the Production of Non alcoholic Sorghum Malt Beverage. *International Journal*

ofFood Science and Nutrition Engineering, 6(4), 81-86.
Gallardo, I., Boffill, Y., Ozuna, Y., Gomez, O., Perez, M., & Saucedo, O. (2013). Production of beverages using malted sorghum сото raw material for celiac patients. *Advances in Science and Engineering,* 4(1), 61-73.
Gallardo, I., Boffill, Y" Rega, L" Pino, M. S" Rodriguez, Y" & Perez, M. (2018). Improving the malting process of udg-110 sorghum in the production of beverages for celiac patients. *CentroAzucar,* 45(2), 46-58.
Gonzales, R., & Ozuna, Y. (2007). *Preliminary study on the production of maltine for celiac children from malted sorghum.* Course Project, Universidad Central Marta Abreu de Las Villas, Santa Clara, Cuba.
Gonzalez, J. M. G. (2013). Process simulation in chemical engineering. *Investigation Cientifica,* 7, 36-42.
Grech, P. (2002). *Introduction to Engineering. An approach through design.* Bogota, Colombia: Prentice Hall.
Igyor, M. A., Ogbonna, A. C., & Palmer, G. H. (2001). Effect of malting temperature and mashing methods on sorghum wort composition and beer flavour. *Process Biochem.,* 36, 1039-1044.
Intelligen (2014). SuperPro DesignerScotch Plains, NJ, USA: Intelligen, Inc.
Jacob, A. A., Fidelis, A. E., Salaudeen, K. O., & Queen, K. R. (2013). Sorghum: Most under-utilized grain of the semi-arid Africa. *Scholarly Journal of Agricultural Science*, 4(3), 147-153.
Jimenez, A. C. (2003). *Diseno de Procesos en Ingenieria Quimica.* Mexico: Reverte S.A.
Kayode, A. P. P., Hounhouigana, J. D., Nout, M. J. R., & Niehof, A. (2007). Household production of sorghum beer in Benin: technological and socio-economic aspects. *Int. J. Consum. Stud.,* 31, 258-264.
Konfo, T. R. C., Adjou, S. E., Dahouenon-Ahoussi, E., Soumanou, M. M., & Sohounhloue, C. K. D. (2014). Physico-chemical profile of malt produced from two sorghum varieties used for local beer (Tchakpalo) production in Benin. *International Journal ofBiosciences,* 5(1), 217-225.
Law, A. M. (2011). *Simulation modeling andanalysis* (3rd ed.). NewYork: McGraw-Hill.
Lyumugabe, F., Gros, J., Nzungize, J., Bajyana, E., &Thonart, P. (2012). Characteristics of African traditional beers brewed with sorghum malt: a review. *Biotechnol. Agron. Soc. Environ.,* 16(4), 509-530.
Maoura, N., & Pourquie, J. (2009). Sorghum beer: production, nutritional value and impact upon human health. In V. R. Preedy (Ed.), *Beer in health disease prevention* (pp. 53-60). Burlington, MA, USA: ElsevierAcademic Press.

Martin, M. (2016). *Industrial Chemical Process Analysis and Design.* Cambridge, United States: Elsevier.

Martinez, V. H. (2012). *Simulacion de procesos en Ingenieria Quimica (*1st ed.). Madrid, Spain: Plaza y Valdez.

MATCHE (2019). Chemical Equipment Cost. Retrieved January 30, 2019, from www.matche.com

Moran, S. (2015). *An Applied Guide to Process and Plant Design.* Oxford, UK: Butterworth-Heinemann.

Morrall, P" Boyd, H. K" Taylor, J. R. N" & Walt, W. H. V. D. (1986). Effect of germination time, temperature and moisture on malting of sorghum (Sorghum bicolor). *J. Inst. Brew.,* 92, 439-445.

Nieblas, C., Gallardo, I., Rodriguez, L., Carvajal, N., Gonzalez, J. F., & Perez, M. (2016). Obtaining beverages and other food products from two sorghum varieties. *Centro Azucar,* 43(3), 66-77.

Okafor, N., & Aniche, G. N. (1980). Brewing a lager beer from Nigerian sorghum. *Brew. Distilling Int.,* 10, 32-35.

Okolo, B. N., & Ezegou, L. I. (1996). Duration of final warm water steep as a crucial factor in protein modification in sorghum malts. *J. Inst. Brew.,* 102, 167-177.

Oramas, G. (2002). Obtaining dual-purpose sorghum (Sorgum bicolor) varieties through the method of progeny selection by furrow. *Agrotecnia de Cuba,* 28(1), 39-48.

Ortega, M. (2016). *Production of beer using sorghum grain integrally.* Diploma thesis, Department of Chemical Engineering, Faculty of Chemistry-Pharmacy, Universidad Central "Marta Abreu" de Las Villas, Santa Clara, Cuba.

Ostrowski, B. (1998). Intensive systems in winter. *Dairy World,* 4(44), 148-155.

Owuama, C. I., & Adeyemo, M. O. (2009). Effect of exogenous enzymes on the sugar content of wort of different sorghum varieties. *World Applied Sciences Journal*, 7, 1392-1394.

Ozuna, Y. (2008). *Obtaining maltine from malted sorghum for celiac children.* Diploma Thesis, Department of Chemical Engineering, Faculty of Chemistry - Pharmacy, Universidad entral "Marta Abreu" de Las Villas, Santa Clara, Cuba.

Pacheco, D. R. (1998). *Agronomic characterization of sixteen improved maicillos* (*Sorghum bicolorL. Moench) in different locations*, El Zamorano, Honduras.

Perez, A., Saucedo, O., Iglesias, J., Wencomo, H. B., Reyes, F., Oquendo, G., & Milian, I. (2010). Characterization and potentialities of grain sorghum (Sorghum bicolor L. Moench). *PastosyForrajes,* 33(1), 1-17.

Perez, E. J. (2016). *Simulation of the beer production process at pilot scale.* Diploma Thesis,

Department of Chemical Engineering, Faculty of Applied Sciences, University of Camaguey, Camaguey, Cuba.

Perry, R. H., & Green, D. (2008). *Chemical Engineers' Handbook.* New York: McGraw- Hill.

Peters, M. S., & Timmerhaus, K. D. (1991). *Plant design and economics for chemical engineers.* NewYork: McGraw-Hill.

Pino, M. S. (2017). *Technology for the production of beer from sorghum malt for celiac patients.* Diploma thesis, Department of Chemical Engineering, Faculty of Chemistry-Pharmacy, Universidad Central "Marta Abreu" de Las Villas, Santa Clara, Cuba.

Ramatoulaye, F., Mady, C., Fallou, S., Amadou, K., Cyril, D., & Massamba, D. (2016). Production and Use Sorghum: A Literature Review. *Journal of Nutritional Health & FoodScience,* 4(1), 1-4.

Ratnavathi, C. V., & Chavan, U. D. (2016). Malting and Brewing of Sorghum *Sorghum Biochemistry:An Industrial perspective* (pp. 63-106). Oxford: Academic Press.

Reyes, S. (2013). *Study of the production of beer from sorghum and barley, at laboratory scale for its implementation in a pilot plant.* Diploma thesis, Universidad Central "Marta Abreu" de Las Villas, Santa Clara, Cuba.

Reyna, L., Robles, R., Reyes, M., Mendoza, Y., & Romero, J. (2004). Enzymatic hydrolysis of starch. *Rev. Per. Qulm. Ing. Qulm.,* 7(1), 40-44.

Rodriguez, L. (2010). *Obtaining bioethanol from sorghum, using the enzymes generated from the malting of the grain itself.* Diploma thesis, Universidad Central "Martha Abreu" de Las Villas, Villa Clara, Cuba.

Rodriguez, L., Gallardo, I., Nieblas, C., Medina, J., & Ortiz, W. (2015). Obtaining dextrinized syrups by enzymatic hydrolysis of sorghum starch. *Centro Azucar,* 42(4), 49-58.

Rodriguez, L., Gallardo, I., Nieblas, C., & Ortiz, W. (2015). Evaluation of two sorghum varieties for starch production. *CentroAzucar,* 42(1), 88-95.

Rooney, L. W., & Serna, S. O. (2000). Sorghum. In K. Kulp & J. Ponte (Eds.), *Handbook of Cereal Science y Technology (2nd* ed., pp. 149-175). New York, USA: Marcel Dekker.

Ruiz-Mercado, G., & Cabezas, H. (2016). *Sustainability in the Design, Synthesis and Analysis Of Chemical Engineering Processes.* Oxford, UK: Butterworth- Heinemann.

Salermo, J. C. (1998). Forage crops at their best. *Mundo Lacteo, 4*(40), 46-58.

Sawadogo-Lingani, H. (2007). The biodiversity of predominant lactic acid bacteria in dolo and pito wort, for production ofsorghum beer. *J. Appl. Microbiol.,* 103, 765-777.

Scenna, N. J. (1999). *Modeling, Simulation and Optimization of Chemical Processes.* Madrid: Marcombo.

Serna, S. S. (1998). Starch refining and production of glucose syrups in a continuous system from different sorghums and corn. Monterrey, N.L., Mexico: ITESM.

Solange, A., Georgette, K., Gilbert, F., Marcellin, D. K., & Bassirou, B. (2014). Review on African traditional cereal beverages. *American Journal of Research Communication,* 2(5), 103-153.

Taylor, J. R. N., & Dewar, J. (1994). Role of alpha-glucosidase in the fermentable sugar composition ofsorghum malt mashes. *J. Inst. Brew.,* 100, 417-419.

Towler, G., & Sinnott, R. (2008). *Chemical Engineering Design. Principles, Practice and Economics of Plant and Process Design.* Burlington, USA: Butterworth- Heinemann.

V alencia, R. C., & Rooney, W. B. L. (2009). Genetic Control of Sorghum Grain Colour. El Salvador: Centro Nacional de Tecnologia Agropecuaria y Forestal (CENTA).

V alle, M. A. M. (2013). Simulation Software for Chemical Engineering. Retrieved November 27, 2018, from www.simulaci6nprocesos.com

V itale, J. D. (1998). Expected effects of devaluation on cereal production in the Sudanian region of Mali. *Agricultural Systems,* 57(4), 489-502.

Annexes

Annex 1. Block diagrams of the production process of maltine from red sorghum CIAP R-132.

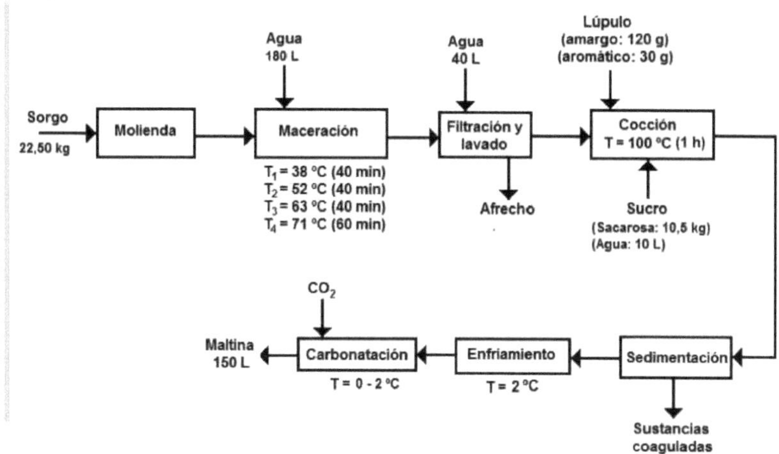

Annex 2. Flow diagrams of the maltine production process from red sorghum CIAP R-132 obtained using the SuperPro Designer' simulator.

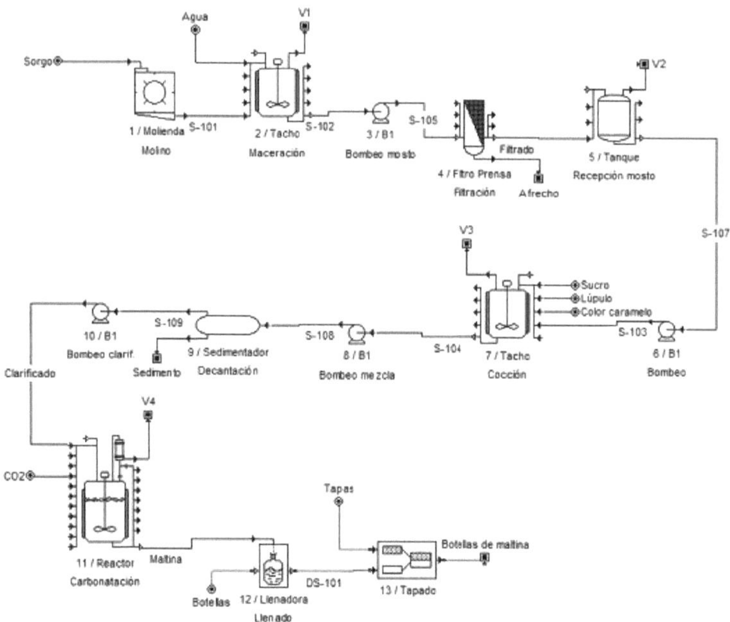

Annex 3. Temperature breaks used during the maceration stage.

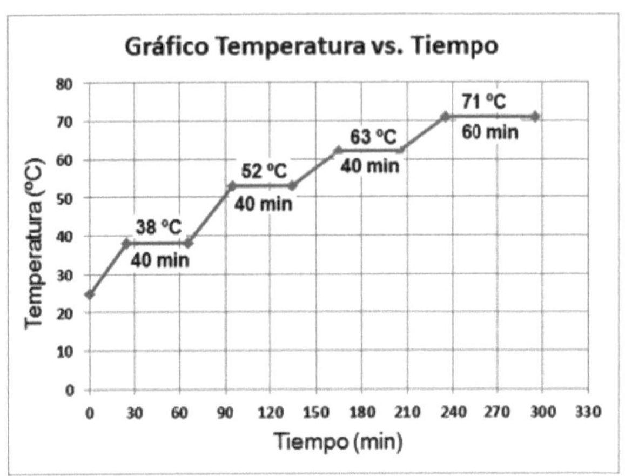

Annex 5. Cost of the main equipment used in the maltine production process from red sorghum CIAP R-132.

Team	Features	Quantity	Total cost
Disc mill	10 kg/h	1	$ 10 200
Macerator	180 L	1	$ 13 000
Filter strainer	12 m^2	1	$3 000
Must reception tank	200 L	1	$ 10 000
Settling tank	180 L	1	$ 14 300
Reactor	200 L	1	$ 15 700
Sanitary centrifugal pump	3,5 kW	1	$4 900
Total			**$71 100**

Annex 6. Percentage chemical composition of sorghum, hops and caramel-coloured hops used in the simulation.

Component	Sorghum -(%)-	Lupulo -(%)-	Caramel colour (%)
Humidity	10,6	10,0	37,5
Starch	69,3	-	-
Protein	12,5	15,0	-
Grease	3,4	-	-
Fibre	2,2	-	-
Ashes	2,0	3,0	0,5
Resins	-	15,0	-
Tannins	-	4,0	-
Alpha-acids	-	8,0	-
Carbohydrates			62,0
Lignin	-	45,0	-

Annex 7. Results of the design of experiments comple

run	*block*	*Capacity*	*Buy sorghum*	*Sale maltina*	*VAN*	*IRR*	*PRI*
1	1	-1,0	1,0	1,0			
2	1	-1,0	1,0	-1,0			
3	1	0,0	-1,0	-1,0			
4	1	0,0	0,0	0,0			
5	1	-1,0	1,0	0,0			

6	1	1,0	-1,0	1,0			
7	1	1,0	-1,0	0,0			
8	1	1,0	0,0	1,0			
9	1	1,0	-1,0	-1,0			
10	1	-1,0	-1,0	1,0			
11	1	0,0	1,0	-1,o			
12	1	0,0	0,0	0,0			
13	1	-1,0	0,0	1,0			
14	1	1,0	1,0	0,0			
15	1	-1,0	-1,0	-1,o			
16	2	0,0	1,0	1,0			
17	2	0,0	0,0	0,0			
18	2	1,0	-1,o	-1,o			
19	2	0,0	0,0	0,0			
20	2	-1,0	1,0	0,0			
21	2	0,0	-1,o	1,0			
22	2	1,0	-1,o	-1,o			
23	2	1,0	-1,o	1,0			
24	2	1,0	0,0	-1,o			
25	2	-1,o	-1,o	0,0			
26	2	1,0	1,0	1,0			
27	2	0,0	0,0	0,0			
28	2	-1,o	-1,o	1,0			
29	2	1,0	1,0	0,0			
30	2	-1,o	0,0	-1,o			

Annex 8.Results of the optimised design of experiments

	run	*block*	*Capacity*	*Buy sorghum*	*Sale maltina*
*	1	1	-1,0	1,0	1,0
*	2	1	-1,0	1,0	-1,0
	3	1	0,0	-1,0	-1,0
	4	1	0,0	0,0	0,0
	5	1	-1,0	1,0	0,0
*	6	1	1,0	-1,0	1,0
	7	1	1,0	-1,0	0,0
	8	1	1,0	0,0	1,0
*	9	1	1,0	-1,0	-1,0
*	10	1	-1,0	-1,0	1,0
	11	1	0,0	1,0	-1,0
	12	1	0,0	0,0	0,0
	13	1	-1,0	0,0	1,0
	14	1	1,0	1,0	0,0
*	15	1	-1,0	-1,0	-1,0
	16	2	0,0	1,0	1,0
	17	2	0,0	0,0	0,0
*	18	2	1,0	-1,o	-1,o
	19	2	0,0	0,0	0,0
	20	2	-1,o	1,0	0,0
	21	2	0,0	-1,o	1,0
*	22	2	1,0	-1,o	-1,o
*	23	2	1,0	-1,o	1,0
	24	2	1,0	0,0	-1,o
	25	2	-1,o	-1,o	0,0
*	26	2	1,0	1,0	1,0

	27	2	0,0	0,0	0,0
*	28	2	-1,o	-1,o	1,0
	29	2	1,0	1,0	0,0
	30	2	-1,o	0,0	-1,o

Annex 9. Values to be presented by each of the 3 input variables considered taking into account a range of variation of ± 20%, y the values to be presented by these 3 parameters in the optimised design of experiments.

Parameter	**Value used in the Variant**	**Minimum value (-20 %)**	**Maximum value (+ 20 %)**
Production capacity (L/batch)	150	120	180
Sorghum purchase cost ($/kg)	14,05	11,24	16,86
Maltina selling price ($/bottle)	6,0	4,8	7,2

Corrida	**Production capacity (L/batch)**	**Sorghum purchase cost ($/kg)**	**Maltina selling price ($/bottle)**
1	120	16,86	7,2
2	120	16,86	4,8
3	180	11,24	7,2
4	180	11,24	4,8
5	120	16,86	7,2
6	120	11,24	4,8
7	180	16,86	4,8
8	180	11,24	4,8
9	180	11,24	7,2
10	180	16,86	7,2
11	120	11,24	7,2

Additional technical-economic results obtained during the simulation of the maltine production process in SuperPro Designer®.

Indicator	***Value***
Total Direct Plant Cost (TDPC)	
Equipment purchase cost	$ 89 000,00
Installation	$ 37 000,00
Pipes	$ 31 000,00
Instrumentation	$ 36 000,00
Isolation	$ 3 000,00
Electrical systems	$ 9 000,00
Buildings	$ 40 000,00
Land improvement	$ 13 000,00
Auxiliary installations	$ 36 000,00
Total CTDP	***$ 294 000,00***
Total Indirect Cost of Plant (TCIP)	
Engineering	$ 73 000,00
Construction	$ 102 000,00
Payments to the contractor	$ 23 000,00
Contingencies	$ 47 000,00
Total CTIP	***$ 245 000,00***
Direct Fixed Capital (DFC) = DFC + PDC + PITC	***$ 539 000,00***
Other	
Working Capital	$ 47 000,00
Start-up cost	$ 27 000,00

Maltine production [bottles/year].	403 460
Gross annual profit [$/year].	$ 915 000,00
Annual net profit [$/year].	$ 600 000,00
Gross margin [%] [%] Gross margin	37,79 %
Annual salary expenditure [$/year] Annual salary expenditure [$/year] Annual salary expenditure [$/year] Annual salary expenditure [$/year	$ 92 000,00
Annual expenditure per raw material [$/year].	$ 403 000,00
Expenditure per expendable material [$/year].	$ 32 000,00
Annual ancillary services consumption expenditure [$/year].	$ 24 000,00
Batch time [h] Batch time [h	15,48 h
Total number of lots/year	387 lots/year

Consumption of raw materials and their percentage influence on the cost of production.

Raw material o material	Annual quantity consumed	Annual cost -($)-	%
Sorghum	17 696 kg	248 624	61,74
Water	168 m^3	17	0,00
Sucrose	9 314 kg	3317	0,82
Lupulo	133 kg	15 505	3,85
Caramel colour	95 hL	5 143	1,28
Carbon dioxide	8 870 kg	3 714	0,92
Bottles	403 460 U	123 459	30,66
Covers	403 460 U	2 824	0,70

Results obtained for each of the experimental runs included in the sensitivity study.

Corrida	*VAN ($)*	*IRR (%)*	*PRI (anos)*
1	3 315 841	59,14	1,11
2	-13311	6,64	7,11
3	7 997 968	103,36	0,51
4	3 027 315	55,39	1,21
5	3 315 841	59,14	1,11
6	358 242	15,23	4,32
7	2 440 215	47,42	1,46
8	3 027 315	55,39	1,21
9	7 997 968	103,36	0,51
10	7 418 723	98,20	0,56
11	3 706 360	64,14	1,00

Net Present Value

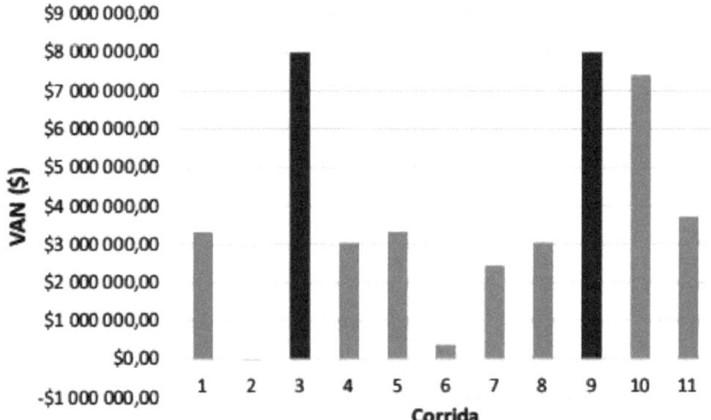

Additional results obtained with respect to the statistical correlation study carried out between the three selected input variables and the indicators NPV, IRR and PRI.

- Net Present Value

Multiple Regression - VAN

Dependent variable: NPV ($)

Independent variables:

Production capacity (L/batch)

Sorghum cost ($/kg)

Maltina sales ($/bottle)

Parameter	Estimate	Standard Error	T Statistic	P-Value
CONSTANT	-1.43956E7	1.48404E6	-9,70028	0,0000
Production capacity	57633,9	5262,99	10,9508	0,0000
Sorghum cost	-99797,7	56188,5	-1,77612	0,1190
Sale maltina	1.77474E6	127449,	13,9251	0,0000

Analysis of Variance

Source	Sum of Squares	Df	Mean Square	F-Ratio	P-Value
Model	7,64337E13	3	2,54779E13	101,45	0,0000
Residual	1.75796E12	7	2.51138E11		
Total (Corr.)	7,81917E13	10			

R-squared = 97,7517 percent

R-squared (adjusted for d.f.) = 96,7882 percent

Standard Error of Est. = 501137,

Mean absolute error = 389393, Durbin-Watson statistic = 2.17855 (P=0.6812) Lag 1 residual autocorrelation = -0.177864

Internal Rate of Return

Multiple Regression - IRR

Dependent variable: IRR (%)

Independent variables:

Production capacity (L/batch)

Sorghum cost ($/kg)

Maltina sales ($/bottle)

Parameter	Estimate	Standard Error	T Statistic	P-Value
CONSTANT	-151,032	3,37928	-44,6935	0,0000
Production capacity	0,659782	0,0119842	55,0542	0,0000
Sorghum cost	-1,14981	0,127945	-8,9867	0,0000
Sale maltina	20,7559	0,290211	71,5201	0,0000

Analysis of Variance

Source	Sum of Squares	Df	Mean Square	F-Ratio	P-Value
Model	10276,3	3	3425,44	2630,57	0,0000
Residual	9,11518	7	1,30217		
Total (Corr.)	10285,4	10			

R-squared = 99.9114 percent

R-squared (adjusted for d.f.) = 99,8734 percent

Standard Error ofEst. = 1,14113

Mean absolute error = 0,835027

Durbin-Watson statistic = 2.27962 (P=0.7439)
Lag 1 residual autocorrelation = -0.203268

Investment Recovery Period

Multiple Regression - PRI

Dependent variable: PRI (years)
Independent variables:
Production capacity (L/batch)
Sorghum cost ($/kg)
Maltina sales ($/bottle)

		Standard	*T*	
Parameter	*Estimate*	*Error*	*Statistic*	*P-Value*
CONSTANT	12,5009	4,08892	3,05725	0,0184
Production capacity	-0,0356912	0,0145009	-2,46131	0,0434
Sorghum cost	0,0874503	0,154814	0,564873	0,5898
Sale maltina	-1,05221	0,351155	-2,99641	0,0200

Analysis of Variance

Source	*Sumof Squares*	*Df*	*Mean Square*	*F-Ratio*	*P-Value*
Model	28,462	3	9,48734	4,98	0,0371
Residual	13,3455	7	1,90651		
Total (Corr.)	41,8076	10			

R-squared = 68.0787 percent
R-squared (adjusted for d.f.) = 54,3981 percent
Standard Error of Est. = 1,38076
Mean absolute error = 0,959251
Durbin-Watson statistic = 1,90502 (P=0,4904)
Lag 1 residual autocorrelation = -0,0050996

Gantt chart obtained using the SuperPro Designer® simulator.

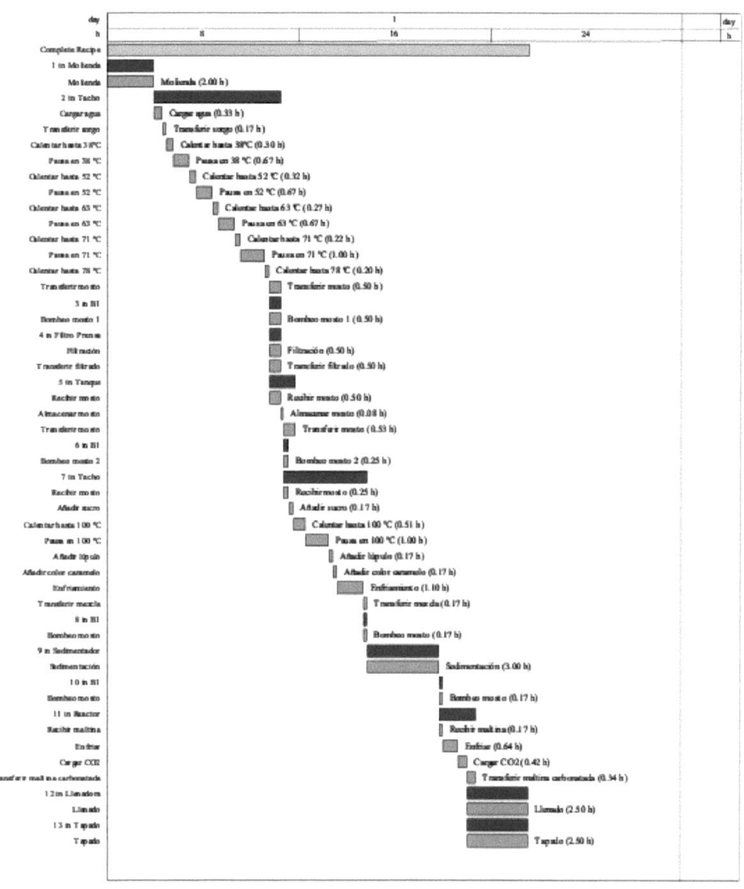

Printed by Books on Demand GmbH, Norderstedt / Germany